Body Needs

PROTEINS

for a healthy body

Heinemann Library
Chicago, Illinois

Angela Royston

© 2003 Heinemann Library
a division of Reed Elsevier Inc.
Chicago, Illinois

Customer Service 888-454-2279

Visit our website at www.heinemannlibrary.com

Created by the publishing team at Heinemann Library
Designed by Ron Kamen and Celia Floyd
Illustrations by Geoff Ward
Originated by Ambassador Litho
Printed in China by Wing King Tong

07 06 05 04 03
10 9 8 7 6 5 4 3 2 1

Library of Congress Cataloging-in-Publication Data
Royston, Angela.
  Proteins for a healthy body / Angela Royston.
     v. cm. -- (Body needs)
Includes bibliographical references and index.
Contents: Why do we need to eat? -- Protein -- What is protein? --
Protein from food -- Vegetable protein -- How your body digests protein
-- Enzymes -- Absorbing -- Using protein -- Proteins on the move --
Growing new cells -- Protein as a source of energy
ISBN 1-4034-0759-2 (lib. bdg.) ISBN 1-4034-3312-7 (pbk.)
  1. Proteins in human nutrition--Juvenile literature. [1. Proteins. 2. Nutrition.] I. Title. II. Series.
  QP551.R867 2003
  612.3'98--dc21

2002012644

Acknowledgments
The author and publishers are grateful to the following for permission to reproduce copyright material: pp. 4, 7, 30, 39, 40, 41 Gareth Boden; pp. 6 (Andrew Syred), 17 (Alred Pasieka), 18, 21, 22 (CNRI), 24 (Volker Steger), 28 (Gusto), 29 (BSIP VEM), 33 (Peter Menzel) SPL; pp. 8, 10, 12, 31, 38 Liz Eddison; pp. 11, 27, 34 Corbis; p. 13 Trevor Clifford; p. 23 Imagebank; p. 26 FPG.

Cover photograph of nuts reproduced with permission of Gareth Boden.

Every effort has been made to contact copyright holders of any material reproduced in this book. Any omissions will be rectified in subsequent printings if notice is given to the publisher.

Some words are shown in bold, **like this.** You can find out what they mean by looking in the glossary.

Contents

Why Do We Need to Eat?

Most people eat three main meals a day. We eat because we get hungry and because we enjoy the taste of food. At the same time, we supply our bodies with the **nutrients** we need to stay alive and healthy.

Cells

Your body is made up of millions of tiny **cells.** Your bones, for example, consist of bone cells. Your skin is made of skin cells, and your muscles are made of muscle cells. Cells are the smallest unit of a plant or animal. Most cells are so small you need a microscope to see them. Each one is like a miniature factory, working hard to carry out a certain task. To do this, your cells need a constant supply of **energy.** They also need many different chemicals, which come mainly from the food you eat. These chemicals are called nutrients.

This family is eating a meal that contains several different kinds of food, including rice, meat, vegetables, bread, and milk. Each of these foods contains a mixture of nutrients.

Nutrients

Carbohydrates, **fats**, proteins, **vitamins**, and **minerals** are all different kinds of nutrients. Most foods contain a lot of one kind of nutrient. But most also contain small amounts of other nutrients, too. Together, nutrients provide energy and materials that the body needs to work properly and to grow. This book is about protein—what it is and how the body uses it. Your body needs other nutrients as well as protein, so we will look at how these other nutrients keep you alive and healthy. We'll also learn how protein does the same thing.

Energy food

Your body's main need is for food that provides energy. Everything you do uses energy. You need energy to run and move around, but you also use energy when you think, eat, and sleep. Carbohydrates and fats provide energy. The body uses carbohydrates, just like a car engine burns gas, and it needs a big supply every day. Foods such as bread, pasta, and potatoes are mainly made of carbohydrates and are your body's main source of energy.

Fats

Fats provide even more energy than carbohydrates. Foods that contain butter, oil, or margarine contain fats. A small amount of fat can supply a lot of energy. However, if you eat more carbohydrates or fats than your body needs for energy, it stores the extra amount as body fat. You need some body fat to keep you warm, but too much can be unhealthy.

Water and Fiber

In addition to nutrients, your body gets two other necessary things from food—water and **fiber**. Health experts recommend that adults get 64 ounces (1.9 liters) of water a day to make up for the liquids they lose. Fiber helps keep your digestive system working well.

The Importance of Protein

Protein is important because you need it to make new **cells.** Each of your cells consists mainly of water and protein. Your body has to repair and replace your cells constantly, and, while you are still growing, it has to create millions of extra cells.

Replacing cells

Most of your cells do not live as long as you do. Skin cells, for example, each last only about 25 days. Your body is making huge numbers of cells all the time to replace those that have worn out and broken down. For example, it makes about 2,300,000 new red blood cells every second! Red blood cells are made in the **bone marrow** at the center of some bones and then released into the blood. Most kinds of cells, however, are made in the place where they are needed. Skin cells are made in the lower layers of the skin, and stomach cells are made in the wall of the stomach.

Through a microscope you can see the dead cells on the surface of your skin. As dead cells flake off, new cells are made to take their place.

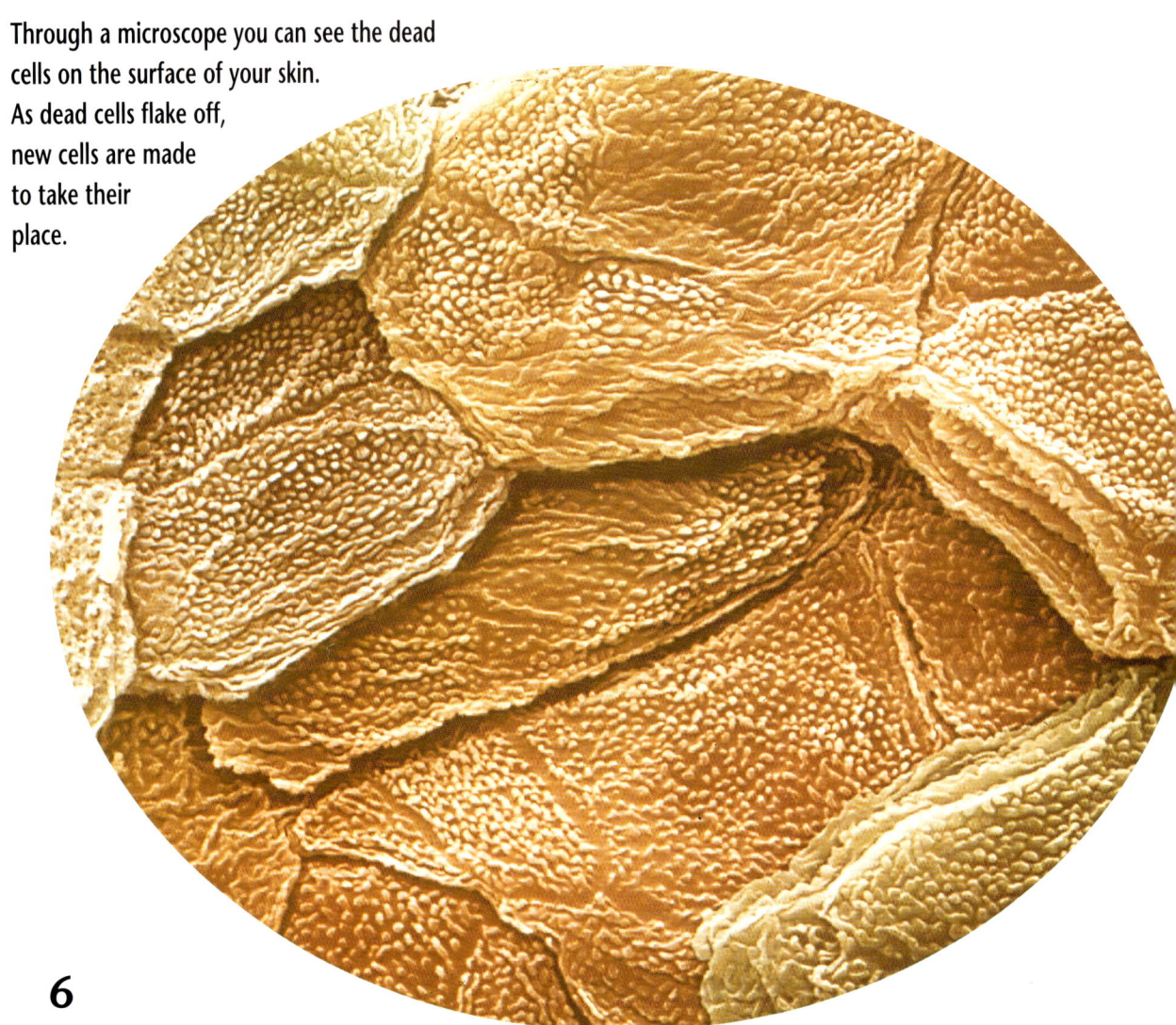

Growing

From the time you are born until you are an adult, your body is constantly growing. It grows so slowly that you do not notice the change from day to day. Even so, while you are a child, your body is having to make millions of extra new cells to produce bigger bones, extra blood, skin, and so on. This means that it is very important for children to eat foods that contain protein.

Protein in food

Meat, fish, eggs, milk, cheese, and beans all contain a lot of protein. Your body cannot use the proteins in the form that you eat them. First it has to break them down into separate units. Then your body rebuilds the units into the particular proteins that it needs. If your food does not give you enough **energy,** your body can use the units of protein to make energy instead.

Vitamins and minerals

Your body needs small amounts of some **vitamins** and **minerals** to help it build new cells. Vitamins and minerals that you get from your food work with protein to build new cells. For example, calcium is a mineral that makes your bones strong. Cells also need certain vitamins and minerals to work properly.

Body Facts

You cannot see new body cells appearing, but they are being made all the time. Cells are so small, that it would take about 100 of them laid end to end to stretch across the head of a pin!

Many people grow fast between the ages of ten and fifteen. This boy has grown out of his pajamas!

What Is Protein?

Protein is one of the fundamental components of living **cells.** There are thousands of different kinds of protein in the human body, in plants, and in animals. Proteins in food provide the raw materials from which all the cells in our bodies are built.

Protein in the body

Much of your body is made of different kinds of protein. Muscles, for example, consist mostly of one kind of protein, while bones contain another kind of protein. Hair, nails, and the outer layer of skin do not look alike, but they are all made of **keratin.** This is a protein that makes them hard and tough. Skin does not feel hard, but it is covered with tough flakes of keratin that make it waterproof. Animals have keratin in their body parts, too. It forms feathers, horns, claws, and scales.

Kinds of protein

Your body creates thousands of different types of protein for all the different parts of your body. The way it makes protein is complex. The body starts by making small protein units called **amino acids.** It does this by putting together substances it takes in from the food you eat. It can combine substances in many different ways. The process of putting together, or combining, substances is called a **chemical reaction.** Scientists refer to the substances themselves as chemicals.

Hair and nails are both made of the protein keratin. Nails are hard, but hair is softened by natural oil, which makes it sleek and bendy.

Amino acids

Amino acids always contain **carbon, hydrogen, nitrogen,** and **oxygen.**
Most also contain **sulfur** and some include **phosphorus,** too. Different
amounts of these elements combine together in different ways to make
various amino acids. Amino acids are the building blocks of proteins.

Chains of amino acids

A protein is made of several different amino acids linked together to
form a chain, kind of like beads in a necklace. Each different kind of
protein consists of a different **sequence** of amino acids. Plants and
animals only use about twenty different kinds of amino acids to form
these chains, but they can be arranged in so many different ways that
they form thousands of different proteins. The beads in a necklace are
linked together by a thread, but in a **molecule** of protein, the amino
acids are held together by **chemical bonds** called **peptide bonds.**

Protein Fact
Each type of protein
contains about 500
different amino acids,
although some have
more and others
have less.

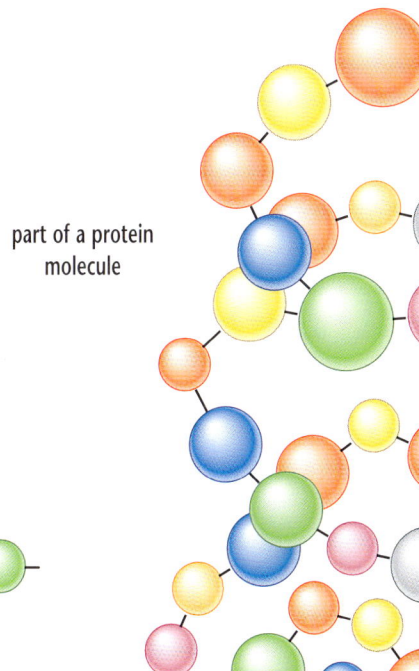

part of a protein
molecule

amino acid

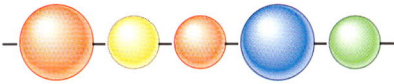

peptide

A protein is a linked chain of amino acids. A simple chain,
or strand, is called a **peptide.** A peptide may be straight
or bent into a three-dimensional shape. A complete
protein molecule contains one or more peptide chains.

Protein from Food

The thousands of different proteins in our bodies are made from just twenty different **amino acids**. The body can make eleven of these amino acids from chemicals in the body, but nine others must come directly from protein-rich food. These nine amino acids are called the **essential amino acids** because we cannot do without them. Some kinds of food contain much more protein than other kinds, and some foods contain all the amino acids we need.

Passing on protein

Plants make all of their own proteins from the **nutrients** they take in from the soil and from water. When cattle, sheep, and other animals eat plants, their bodies change the plant proteins into animal proteins. This means that food from plants and from animals both contain protein. The muscles of animals, however, are particularly rich in essential amino acids. Meat from animals contains all of the essential amino acids our bodies must have, so foods from animals are said to be a complete source of protein for humans.

These foods are all rich in animal protein. They contain all the amino acids that humans need.

Meat

Most of the meat we eat comes from different animals' muscles. When you eat chicken breasts, for example, you are eating the large muscles the bird used to flap its wings. The meat on the legs (drumsticks) is the muscle it used for walking. Steaks and hamburgers come from the muscles of cattle. Pork chops and pork roast come from the muscles of pigs. Other parts of animals contain protein, too. For example, some people eat chicken or beef **livers.** Hot dogs made of pork or beef muscle often also contain small amounts of heart, liver, or kidney.

Seafood

Seafood includes fish and shellfish. White fish contains less **fat** than **red meat** and is a good source of protein. The white or colored flesh of fish, such as cod, haddock, tuna, salmon, and sardines, are the muscles that lie along each side of the fish's backbones. When you eat shrimp, lobster, and other shellfish, you are eating their muscles, too.

Dairy products

Eggs and milk are also a good source of animal protein. Egg white is pure protein, while the yolk is a mixture of fat, protein, and vitamins. The milk we drink comes mainly from cows, but cows' milk is also made into cheese, yogurt, and butter. Some yogurt and cheese is made from sheep or goats' milk.

Cows spend much of the day and night eating grass. Their bodies turn the proteins in grass into animal proteins of different kinds. They pass those proteins on to us in the form of milk and meat.

Vegetable Protein

Although plant protein is called vegetable protein, it comes mainly from plants that we do not think of as vegetables, such as nuts, grains, and **legumes.** Nuts not only contain protein but also many **vitamins, minerals,** and **fats.**

These foods are all rich in protein. For most people, about one third of the protein in their diets comes directly from plants.

Grains

Grains are a source of **carbohydrates,** but they contain vegetable protein, too. Grains include wheat, rice, corn, and oats. Wheat is ground into flour and made into bread, cakes, and pasta. Breakfast cereals, too, are made from grains. They are another source of vegetable protein.

Beans, peas, and lentils

Different kinds of beans, peas, and lentils are known as legumes, and they contain plenty of vegetable protein. In India, a spicy lentil dish known as dal is very popular. Hummus dip, a dish originally from the Middle East, is made from chickpeas. Legumes, including lentils, navy beans, and black beans, are often used in soups. Soybeans, another legume, are very rich in protein.

Soybeans

People who do not eat meat, called vegetarians, often eat soybeans and soybean products. Many of the veggie burgers you can buy at the supermarket are made from soybeans. People in Malaysia, Thailand, and Japan have used soybeans for a long time. They use them to make foods such as tofu and miso soup, which are full of protein. These foods are also quite popular in the United States.

New proteins

You can buy many frozen vegetarian meals in supermarkets and health food stores. One of the newer types of meat substitutes is mycoprotein. Mycoprotein can be grown from **yeast**, other **fungi**, or **bacteria**. The mycoprotein is **fermented.** It is then cooked and flavor is added. It can be flavored to taste like meat, although most vegetarians prefer to add herbs and spices instead of meat flavors.

A complete meal

Even plants that are rich in protein lack one or more of the necessary **amino acids** we need to get from food. But this is not a problem. If you eat different kinds of vegetable proteins, you will get all the different amino acids you need. Many snacks and meals you probably already eat contain a mixture. A snack of peanut butter and crackers mixes two sources of vegetable proteins. A grilled cheese sandwich mixes animal and vegetable proteins, as does breakfast cereal and milk.

Peanuts

Peanuts are not really nuts. They are a type of legume related to peas. Ounce for ounce, peanuts contain more protein than raw steak and more energy than chocolate cookies.

Peanut butter sandwiches are a good source of protein. Both the peanut butter and the bread contain protein, and together they give you all the amino acids you need.

13

How Your Body Digests Protein

Your body cannot use proteins in food in the form that you eat them. This is because they are not the exact proteins that your body needs. They contain the right **amino acids,** but in the wrong amounts and in the wrong **sequences.** Food also has to be digested before your body can use the **nutrients** in it. During digestion the food is broken down into tiny pieces that are small enough to pass through the walls of your intestine into your blood. Protein is broken down into **peptides** and then into separate amino acids.

Cooking

Most of the protein you eat has been cooked. Cooked food is usually softer and easier to chew than raw food. Cooking also kills germs in meat and fish, making them safer to eat.

Chewing

Digestion begins in the mouth, when you crunch and chew your food to break it up. As you chew, **saliva** mixes with the food to form a mushy lump. When the lump is soft enough, you swallow and the food passes down the **esophagus** into your stomach.

In the stomach

The muscles of the stomach act like a slow-moving blender, churning the food around. The stomach juices mix with the food and, together with the churning, turn the chewed-up food into a kind of thick soup, called **chyme.** Food usually stays in the stomach for two to four hours. But some foods that are high in **carbohydrates,** such as spaghetti and rice, pass through faster. Food with a lot of **fats** and protein in it, however, stays in the stomach longer.

Stomach juices

The stomach contains a strong **acid** that attacks the food and kills the germs in it. The acid in the stomach is called gastric juice. It also contains a substance called **pepsin** that breaks up the protein **molecules** into smaller bits.

In the intestines

The chyme passes slowly from the stomach into the small intestine.
Juices from the **liver** and **pancreas** pour into the small intestine.
The intestines make their own digestive juices, too. All these juices work
on proteins, carbohydrates, and fats, breaking them into smaller units.
The juice from the liver is greenish-brown and is called bile. It breaks up
fat into tiny drops.

Breaking down proteins

Chemicals in the juice from the pancreas work on the protein molecules,
breaking them down into smaller and smaller bits. Eventually, the
smallest proteins are broken into separate amino acids.

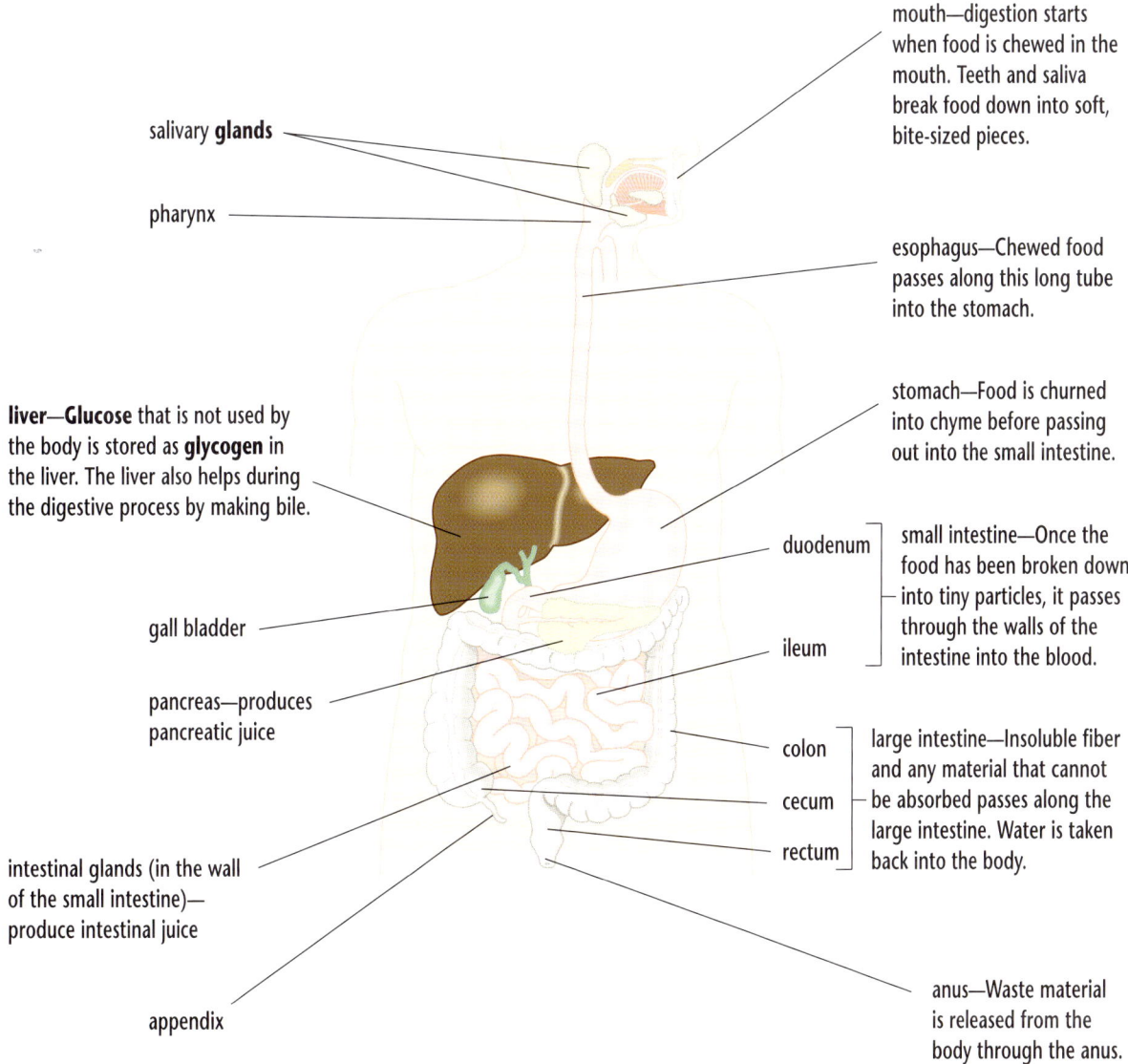

salivary **glands**

pharynx

liver—**Glucose** that is not used by
the body is stored as **glycogen** in
the liver. The liver also helps during
the digestive process by making bile.

gall bladder

pancreas—produces
pancreatic juice

intestinal glands (in the wall
of the small intestine)—
produce intestinal juice

appendix

mouth—digestion starts
when food is chewed in the
mouth. Teeth and saliva
break food down into soft,
bite-sized pieces.

esophagus—Chewed food
passes along this long tube
into the stomach.

stomach—Food is churned
into chyme before passing
out into the small intestine.

duodenum

ileum

small intestine—Once the
food has been broken down
into tiny particles, it passes
through the walls of the
intestine into the blood.

colon

cecum

rectum

large intestine—Insoluble fiber
and any material that cannot
be absorbed passes along the
large intestine. Water is taken
back into the body.

anus—Waste material
is released from the
body through the anus.

Enzymes

Enzymes are special types of proteins. They are found in digestive juices. They help **chemical reactions** take place more quickly, but they are not changed by the reactions. Without different enzymes, your body would not be able to digest food.

How enzymes work

Enzymes break down complex substances into more basic ones. An enzyme attaches itself like a key in a lock to a particular group of chemicals. In digestion, enzymes break down carbohydrates, fats, and proteins into simpler chemicals. Then, the enzymes attach to another similar group of chemicals. Enzymes work very fast. A single enzyme can split a million different chemicals one after the other in just one minute.

Digestive Juices
An adult produces about 10.5 quarts (10 liters) of digestive juices every day. About 1 quart (0.9 liters) is saliva and about 3 quarts (2.8 liters) is gastric juice.

The right conditions

Digestive enzymes only work in the right conditions. The temperature, for example, must be less than 104 °F (40 °C). The temperature of the body is usually around 98.6 °F (37 °C). This temperature is just right for enzymes. Outside the body, food needs to be heated to about 212 °F (100 °C) before it begins to break down. Enzymes allow us to digest food at a much lower temperature.

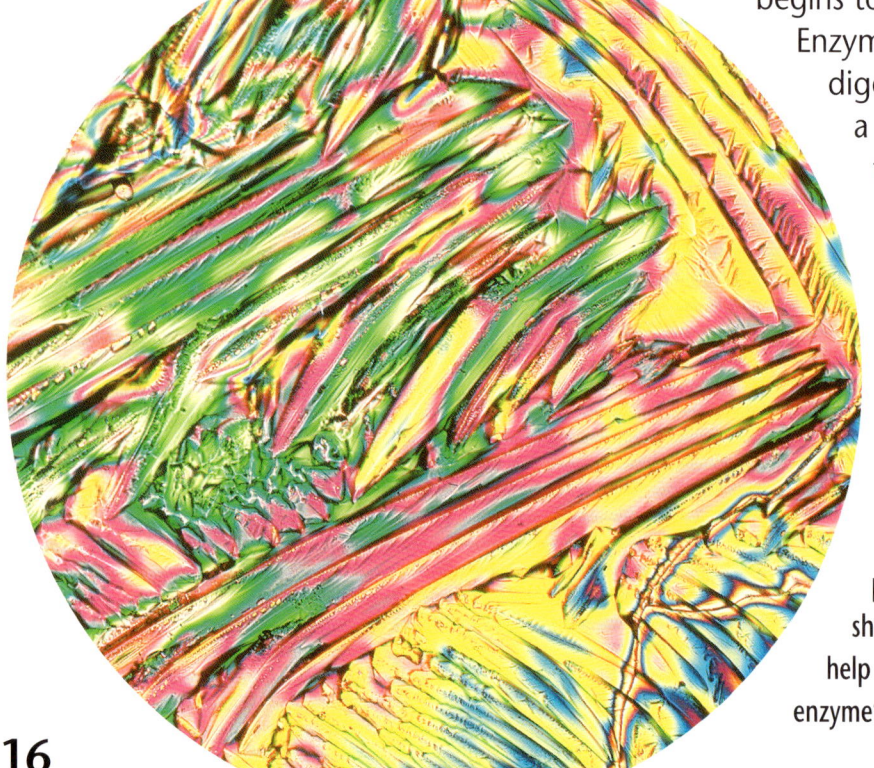

Trypsin, an enzyme that helps digest protein, is shown magnified here. It is one of the two enzymes in the digestive juices made by the pancreas. The colors shown are dyes that help scientists see the enzyme's parts clearer.

Pepsin

When you start to chew, your mouth produces extra **saliva.** At the same time, your stomach produces gastric juices. These juices contain the enzyme called **pepsin,** which is mixed with **hydrochloric acid.** Pepsin starts to work on the protein, breaking it into smaller units of **amino acids.**

Enzymes from the pancreas

The digestive juice produced by the **pancreas** contains several enzymes. Two of these enzymes work on proteins in the small intestine. They break protein into simpler chemicals. They do this by cutting the **peptide bonds** that hold the amino acids together in a chain. These short strands in the chain are finally broken into single amino acids.

Babies

When a baby is born, its digestive system has to start working for the first time. It drinks milk, either from its mother or from a bottle. Babies' stomachs produce an extra enzyme called rennin. It changes the protein in milk into a solid called casein. This solid takes longer to pass through the intestines. This gives the other enzymes more time to work on breaking it down.

This diagram shows how enzymes break up protein. In the stomach some of the long chains of amino acids are cut into shorter strands called peptides. In the small intestine, the peptides are broken into single amino acids.

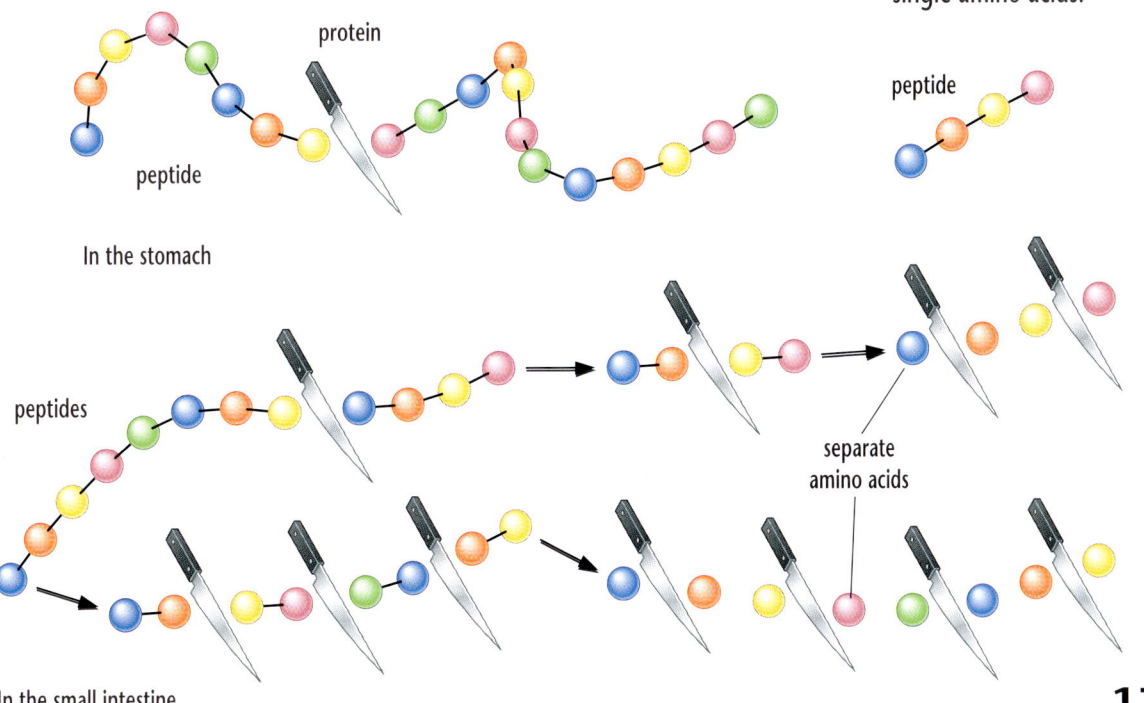

protein

peptide

peptide

peptide

In the stomach

peptides

separate amino acids

In the small intestine

Absorbing Food

Once the protein has been broken down into separate **amino acids,** the particles pass right through the walls of the small intestine into the blood. The blood takes all the particles of digested food straight to the **liver.** The liver is an amazing **organ** that controls hundreds of different **chemical reactions.** One of the liver's many jobs is to check the level of **nutrients** in the blood, releasing fresh supplies as they are needed.

This magnified photo shows some of the villi in the small intestine. Nutrients pass through the walls of the villi into the blood.

Through the intestine walls

The inside lining of the small intestine is covered with tiny fingerlike extensions called **villi.** There are millions of villi on the inside of the small intestine. The villi are about 0.02 inches (0.5 millimeters) long and have even smaller "microvilli" covering them. The villi give the intestine a large surface for digested food to pass into the blood. Their walls are so thin that digested food passes through them into the blood vessels inside.

Body Facts

The liver is the largest organ inside the body. When you are resting, about a quarter of your blood is held in the liver. Here, it is cleaned and processed. Poisons are removed and destroyed and the blood is resupplied with nutrients.

The liver

A vein takes the blood, along with amino acids and other particles of digested food, straight from the small intestine to the liver. The liver uses some of these nutrients to resupply the blood. Your liver stores some extra nutrients, such as sugar, but it cannot store extra amino acids. Instead, it changes some into sugar and some into other amino acids. The liver then changes any leftover amino acids into a waste substance called **urea**. Urea dissolves in water to form urine, which is passed out of the body.

How nutrients reach your cells

Blood, freshly supplied with nutrients, leaves the liver and returns to the heart. The heart sends it to the lungs to pick up **oxygen** and then pumps it through your **arteries** to all the **cells** in your body.

Undigested food

Not all protein in your food is digested. Some of it gets to the end of the small intestine before it has been broken down into small enough strands to pass into the walls of the small intestine. These larger strands pass into the large intestine with other undigested food and slowly make their way to the rectum, from where they leave the body as solid waste.

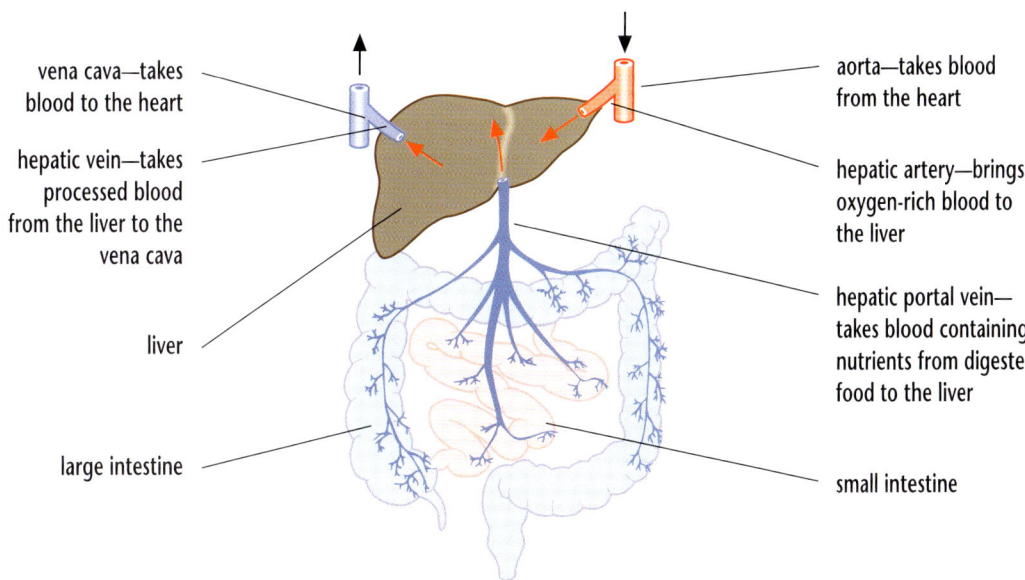

vena cava—takes blood to the heart

hepatic vein—takes processed blood from the liver to the vena cava

liver

large intestine

aorta—takes blood from the heart

hepatic artery—brings oxygen-rich blood to the liver

hepatic portal vein— takes blood containing nutrients from digested food to the liver

small intestine

The liver lies just below your ribs on the right-hand side of your body. As blood from the intestines passes through the liver, the nutrients in the blood are processed.

Using Protein

The **cells** in the body use protein in different ways. Each cell takes the **amino acids** it needs from the blood and uses them to build up new proteins. The cell then uses these new proteins to build new cells or to help it carry out its job in the body.

Protein in the cells

Each cell is made up of several special parts. Some parts of the cell deal with protein. Every cell has **ribosomes.** Ribosomes are microscopic factories that make proteins. To do this, they join together the right amino acids in the right order. Different kinds of cells make different proteins in their ribosomes. The proteins are then stored in a part of the cell called the Golgi complex. This strange-shaped structure is a kind of warehouse, where the proteins are kept until they are needed. A thin covering, called the cell membrane, surrounds the cell. This membrane is made of protein and **fat.**

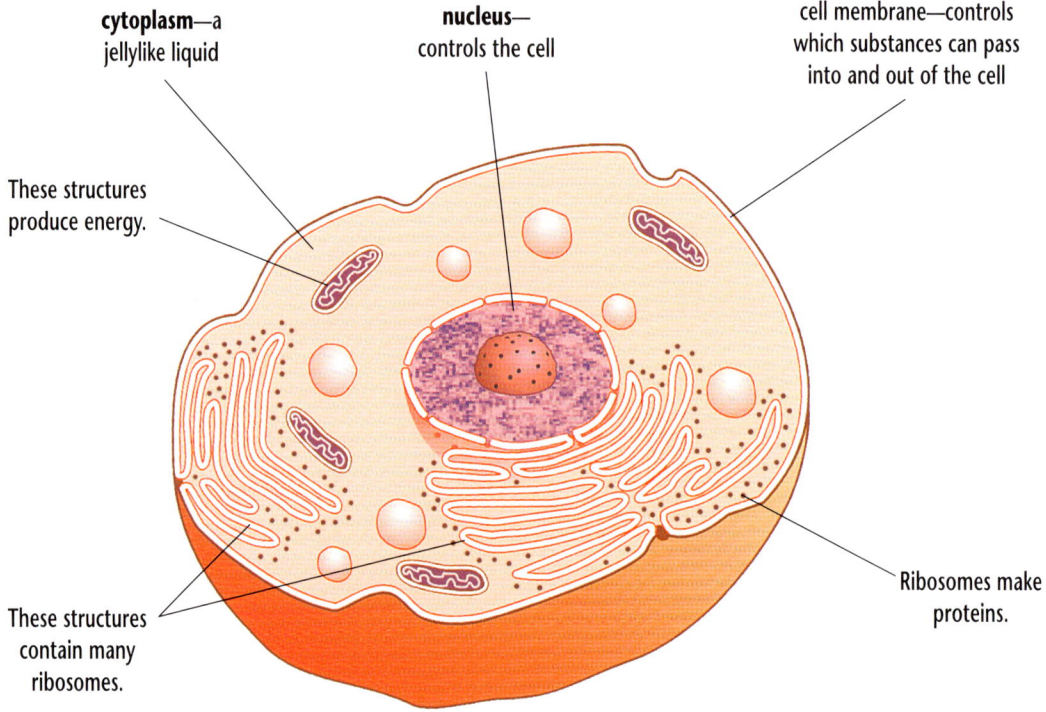

cytoplasm—a jellylike liquid

nucleus— controls the cell

cell membrane—controls which substances can pass into and out of the cell

These structures produce energy.

These structures contain many ribosomes.

Ribosomes make proteins.

Each cell is like a factory with several special units. One kind of unit makes protein that is stored in a different kind of unit.

20

Hemoglobin

Protein plays a key role in helping your body work properly. Red blood cells, for example, contain a protein called hemoglobin. In the lungs, the hemoglobin in red blood cells picks up **oxygen** from the air you have breathed. The oxygen turns the hemoglobin bright red. This bright red blood is pumped by the heart through the **arteries** to different parts of the body. The arteries branch into thinner and thinner tubes called capillaries. The thinnest tubes are so narrow the red blood cells have to change shape to squeeze through them. Here, oxygen moves from the hemoglobin through the wall of the capillary and into the cells that need it. As the blood loses its oxygen, it becomes a darker red. Dark red blood returns through the veins to the heart and lungs, where it picks up a fresh supply of oxygen to deliver more cells.

Collagen

Many parts of your body contain the protein collagen, which is very strong but also very flexible, or bendy. Collagen fills the spaces between the cells in bones and other **tissues** such as ligaments and tendons. Ligaments are stretchy bands that hold a joint together but allow it to move. Tendons are similar bands of tissue that attach the muscles to the bones. Collagen is also found in the skin and in the walls of blood vessels. Blood vessels have to be very strong and able to stretch as the heart pumps spurts of blood through them.

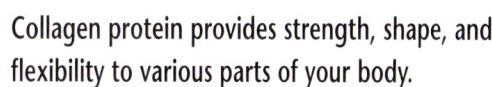

Collagen protein provides strength, shape, and flexibility to various parts of your body.

Protein Fact

When collagen is boiled, it produces a substance called gelatin. Gelatin made from animal collagen is an ingredient in glues, beauty products, and medicines.

Proteins on the Move

Proteins made in some kinds of **cells** are released from the cells to do their work elsewhere. Some proteins, such as **antibodies** and **hormones**, are carried by the blood to other parts of the body.

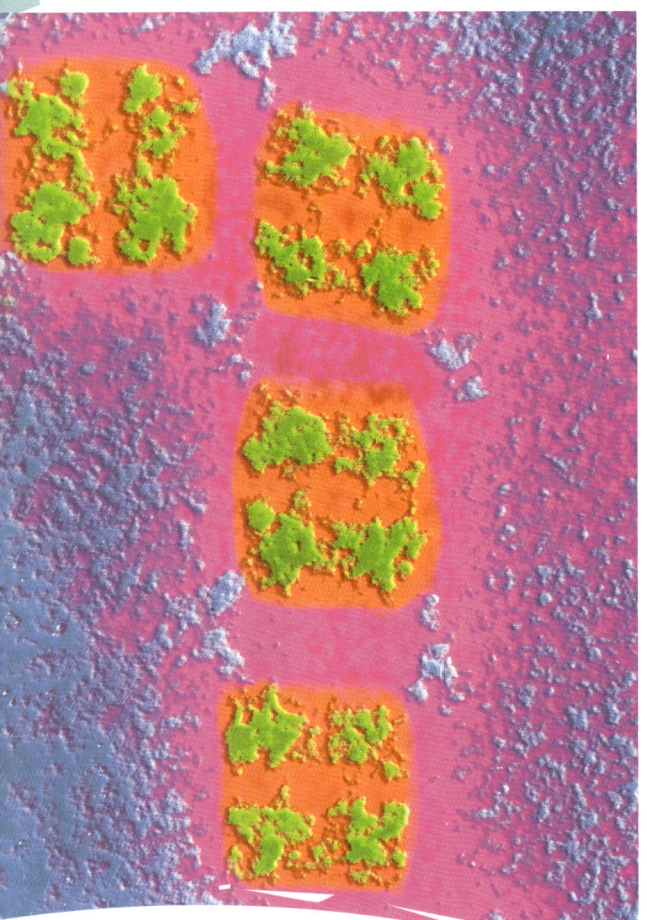

This is what an antibody looks like under a microscope. The colors in the image help scientists clearly see each part of the antibody.

Antibodies

Antibodies are proteins that help the body fight off infection. Infection is a disease caused by **bacteria, viruses,** or other things that are able to enter the human body. The antibodies are made by cells in the **bone marrow,** which is the red, jellylike substance at the center of some bones. Each kind of antibody is especially designed to attack certain kinds of bacteria, viruses, or poisons. Some antibodies attack viruses. They latch on to a virus and make it easier for other blood cells to kill it. Other antibodies deal with bacteria. Even when the infection is over, some of the antibodies remain in the blood. This means that if the same viruses or bacteria get into the body again, the antibodies are ready to attack them immediately.

Hormones

Hormones are proteins that work as chemical messengers. They trigger various processes in the body. Each type of hormone is made in a particular **gland.** For example, the hormone adrenalin is made by two glands just above the kidneys. When you are scared or alarmed, the glands produce extra adrenalin. Adrenalin prepares the body for action. It makes your heart pump faster and your breathing speed up. It also gets the **liver** to release more **glucose** into the blood. Other hormones work over a long period of time. Growth hormone, produced by a gland in the brain, sends the messages that make you grow taller.

Making enzymes

Enzymes are also made of protein. Different enzymes are used all over the body to speed up different **chemical reactions**. Enzymes in the cells help build proteins. Other enzymes in the cells help turn glucose into **energy**. Some enzymes are used in digestion. The cells of the **pancreas**, for example, produce two enzymes that break up protein. They are called trypsin and chymotrypsin.

Enzyme Facts

Trypsin is one of the enzymes that breaks up protein. It is made in the pancreas, but it is so active it could digest the pancreas itself. To protect itself, the pancreas actually produces a substance called trypsinogen. This only turns into trypsin when it reaches the small intestine.

The hormone adrenalin helps athletes perform well. This protein makes them more alert and sends more blood and oxygen to the muscles.

Growing New Cells

Amino acids are used to grow new **cells**, either to create extra ones while you are still growing or to replace cells that have died. Most of your cells do not live as long as you do. After a while they die and have to be replaced by new cells. In addition, some cells may need to be replaced because they have been damaged.

Repairing damaged tissue

When you scrape or cut yourself, some of your cells are destroyed and blood leaks out through the broken skin. Blood contains a protein called fibrinogen. When the blood leaks out and air reaches it, the fibrinogen changes into a solid called fibrin that forms crisscross threads over the wound. The threads slow down the flow of blood, until the blood clots and hardens to form a scab. Under the scab, new cells grow to replace the damaged skin cells. Cells inside your body can become damaged, too. **Bacteria** and **viruses**, for example, damage cells that then have to be replaced.

Growth

Children and teenagers produce the most new cells. However, the most dramatic rate of growth takes place before a human baby is born. One cell grows into a baby weighing about 7 pounds (3 kilograms) and measuring about 20 inches (50 centimeters) long. By the time babies are two years old, they have reached about half their adult height. Growth then slows down but continues, sometimes with faster spurts.

The fibrin cells in this microscopic photo of a wound are colored to be viewed easier. Fibrin helps to seal the wound while new cells are made to replace the damaged ones.

Growth Facts

Every human life begins as one cell. This cell is formed when a human female egg is fertilized by a human male sperm. This single cell divides over and over again, forming the different parts of the body. By the time you are an adult, that one cell has multiplied into at least 50 trillion cells!

Puberty

The time when your body begins to change from a child's into an adult's is known as **puberty**. It usually starts around the age of eleven or twelve and lasts until about the mid-teens, but this varies from one person to another and differs in girls and boys. During this time, your height will increase rapidly. You go on growing more slowly until you reach your full height in your late teens or early twenties. Even then your body still makes extra cells.

How new cells form

New cells are formed by a process called **mitosis**. The **nucleus** in a cell makes an exact copy of itself. The **cytoplasm** inside the cell then divides and forms two cells, each with its own nucleus.

Your body needs to grow new cells as you grow up.

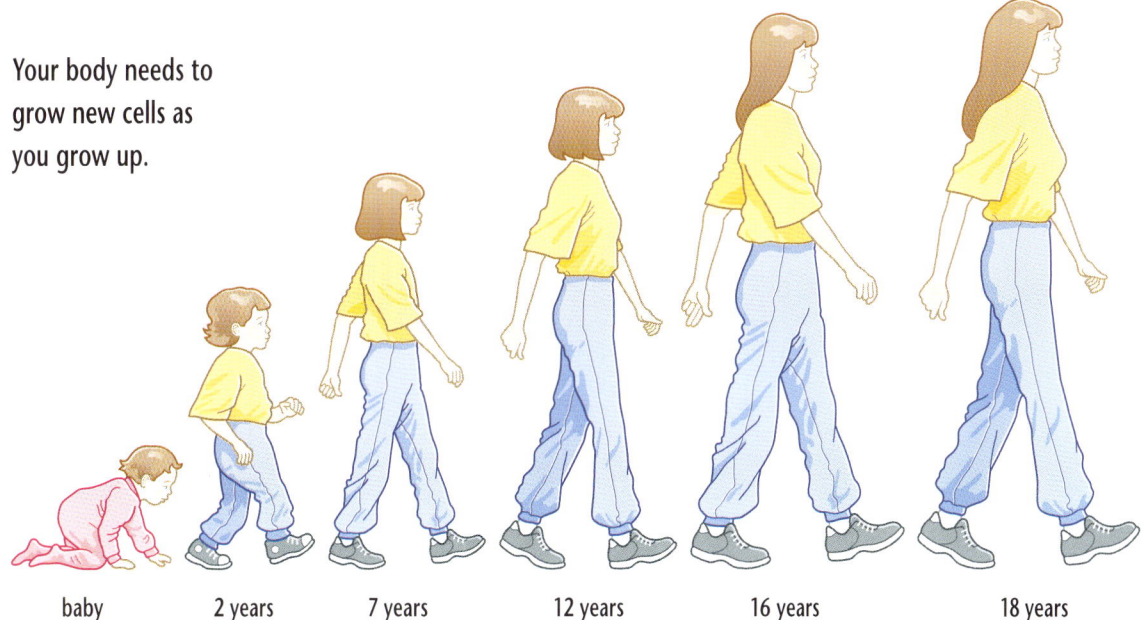

| baby | 2 years | 7 years | 12 years | 16 years | 18 years |

Protein as a Source of Energy

If the blood contains more **amino acids** than the **cells** need, the **liver** converts some of them into **glucose**. Glucose is the sugar that provides the **energy** the cells need in order to work the right way. The glucose may then be released into the blood to supply the cells. Or it may be changed again, this time into a substance called **glycogen.** This substance can be stored in the liver until it is needed.

Energy first

It is important to eat a good balance of energy-rich **carbohydrates** (for example, whole grain breads, rice, pasta, fruits, and vegetables) and proteins. If the food you eat does not provide your body with enough energy, then your liver changes amino acids into glucose instead. The reason it does this is because the body needs energy more than it needs protein. In some cases, so much protein may be used to supply energy, it leaves the body short of **essential amino acids.**

Some people eat very few carbohydrates because they are trying to stay thin. However, when there is not enough glucose in the blood, the liver changes protein into sugar to provide energy instead. When this happens, the body may not get enough needed proteins.

Athletes and body builders

Athletes may spend several hours every day exercising. The exercise strengthens their muscles and makes their heart and lungs grow larger and work better. Some sports demand a lot of strength or endurance, which is the ability to keep exercising for a long time. Weightlifting, for example, demands strength, while running a long race demands endurance. Athletes who take part in these sports often eat a lot of extra protein, but their diets probably include more than enough protein already. If the extra protein is not needed to build muscle, it will be broken down and used to supply energy instead.

Waste protein

The body can store extra carbohydrates and **fats** to use when it needs them, but it cannot store protein. If you eat more protein than your body needs, the liver will change some of the amino acids into glucose or glycogen, but the rest is lost as waste.

As she exercises, this woman's muscles use some extra protein to become thicker and stronger. Weightlifting develops muscles.

Too Much Protein

All the **cells** in the body need protein so they can keep healthy. It is also needed to make new cells. But is it possible to eat too much of this important body-building substance? The answer is yes. When you consume more energy than you use up, the body stores the excess energy as **fat.** Too much food, including too much protein, will make you overweight.

The body stores extra carbohydrates, fats, and protein as fat, so eating too much fast food can make you overweight.

Protein waste

The **liver** processes protein to release the **amino acids** the body needs. If there is extra protein, it changes some into **glucose.** Then the liver changes any remaining amino acids into a waste substance called **urea.** The process of changing amino acids into glucose produces waste **nitrogen.** This is also changed into urea. As the cells use amino acids, they also produce nitrogen as a waste product. The blood carries it to the liver. It is then changed into urea as well.

The kidneys

You have two kidneys, one on each side of the spine, just above your waist. As blood passes through them, the kidneys filter the blood and remove urea and any extra water and salt. Urea is a solid, but it dissolves easily in water to form urine. Urine trickles from the kidneys down two tubes to the bladder. Urine is stored in the bladder. The bladder gradually stretches as it fills up. It can hold up to about half a quart (600 milliliters) of urine at a time.

Extra protein

People who eat too much protein may damage their liver and kidneys. This is because having to deal with the excess protein makes these organs work too hard. The liver has to work hard to change all this extra protein into urea. The kidneys have to work hard to filter it from the blood.

This X-ray shows different parts of the human body, including the kidneys, bladder, and ureters, which are the tubes that carry urine to the bladder. The kidneys process about two pints of blood every minute. The urea and water is filtered out to make urine, which is stored in the bladder.

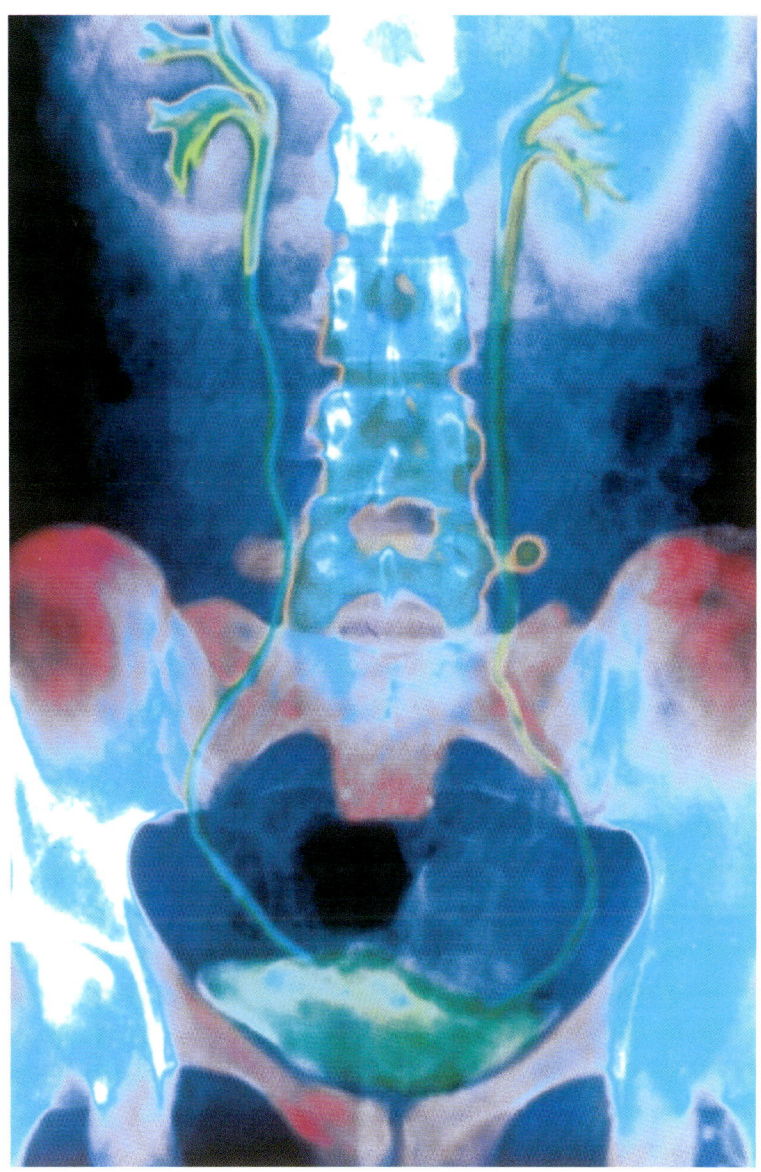

Urea

Nitrogen is an important substance that plants must have to grow. Urea is rich in nitrogen. Artificial, or not naturally made, urea is used as a **fertilizer** in gardens and on crops. Urine, however, does not make a good fertilizer since it is too **acidic**. Artificial urea is also used in cattle feed, plastic, and in medicines.

Food Allergies

A small number of people react to a particular protein in their food. These are people who are allergic to certain substances, such as the protein in cow's milk, wheat, or nuts.

Allergies to protein

When a person is allergic to a particular kind of food, the body acts as if that food were a germ or a poison. Peanuts, milk, cheese, eggs, and wheat are the most common cause of food allergy. Food allergies often cause a person to throw up or to develop an itchy rash. Some people may also have trouble breathing. The best way to treat an allergy is to avoid the food that causes it. People who are allergic to peanuts, for example, need to check the list of ingredients on a label to see if a food contains peanuts or peanut oil. However, even if the ingredients do not list peanuts, there might still be a problem. If a food was manufactured in a factory that processes peanuts, the food might still trigger a reaction. Many food labels now carry warnings to alert people with peanut allergies or other food allergies.

Nut Allergies

When a person who is allergic to nuts eats anything that contains nuts, even a small amount of nuts, his or her body immediately reacts. Within a few minutes, the person's mouth could swell up. The person's skin also might break out in a red, blotchy rash that is very itchy. These symptoms can last for a few hours.

If you are allergic to milk, you should read the ingredients listed on food labels to see if the food contains milk.

Gluten

Gluten is a mixture of two proteins and is what makes bread dough stretchy and bouncy. Bread companies sometimes add extra gluten to the dough to make it softer and lighter. Some people's digestive systems cannot handle gluten in the food they eat. In many cases this is not due so much to an allergy but to the fact that gluten damages **villi** in the small intestine. The gluten wears away the villi so that other **nutrients** may not be absorbed either.

Celiac disease

People who cannot tolerate gluten have celiac disease. They cannot eat foods that contain gluten. Wheat has more gluten than other grains, but rye, oats, and barley contain some, too. Rice and corn are gluten-free, so people with celiac disease can eat foods made with these grains.

All of these foods contain wheat. People who suffer from celiac disease cannot eat any of them.

Extreme Shock

Some people are so allergic to a particular food their bodies go into an extreme shock called **anaphylactic shock** if they eat it. Their blood pressure drops and they pass out. Unless they get medical help at once, they will die. Foods that are most likely to cause anaphylactic shock are nuts, shellfish, and eggs.

Too Little Protein

Most people in **developed countries** such as the United States or the United Kingdom do not have trouble getting enough protein. In fact, most people in developed countries eat more protein than they need. Some people who have eating disorders, such as **anorexia** or **bulimia**, or those who eat only junk food, such as candy, cookies, and potato chips, are likely to get too little protein. However, most people who suffer from a lack of protein live in **developing countries.**

A lack of protein

Many people in developing countries do not have enough money to buy meat, eggs, or other types of animal protein. Children and babies are most affected by not getting enough protein. It affects them in several ways. One of the first signs is that they cannot fight off illness very well. Their bodies are also less able to replace worn out or damaged **cells,** so it takes longer for cuts and wounds to heal. They may feel sick all the time and have trouble learning. They do not grow as tall as they should, because their bodies cannot build all the extra cells needed for growth.

Kwashiorkor

A severe lack of protein leads to a disease called kwashiorkor. It mainly affects young children who have stopped getting milk from their mother. Milk is full of protein. Often, the children are eating solid foods that do not have much protein in them. In the villages and countryside of much of Africa, the Caribbean, and the Pacific Islands, babies as well as adults eat yams, cassavas, sweet potatoes, and green bananas. These foods are rich in **carbohydrates** but low in protein.

Symptoms of kwashiorkor

One of the most obvious signs of kwashiorkor is a large, swollen belly. The swelling can make the baby or child look quite fat. But the swelling is not caused by fat, but by liquid that has collected below the skin. The **liver** also becomes larger. Skin also becomes flaky and hair gets thinner and loses some of its color. The baby or child stops growing and becomes tired and weak.

Treatment of Kwashiorkor

On average, children who are one to three years old are most at risk to develop kwashiorkor. The disease can be deadly if not treated. Treatment usually involves giving the affected child dried skim milk and high-protein foods. However, because the child has most likely not eaten for a long period of time, foods must be reintroduced slowly. **Carbohydrates** are usually given to the child first, followed by protein foods. Even with successful treatment, however, a child who had kwashiorkor is unlikely to develop to a normal physical height or weight.

This young child is suffering from kwashiorkor, caused by a severe lack of protein. Children with kwashiorkor will die unless they are treated.

Famine and Starvation

Famine is the most extreme case of lack of food. People starve when they do not have enough food to supply them with the **energy** they need or with proteins. Their bodies become extremely thin and their **organs** stop working properly. This condition is called marasmus. People with marasmus are starving and will die if they do not get food.

Symptoms of marasmus

People who are starving feel hungry all the time. This is not the sort of hunger that you feel when your stomach rumbles and you are looking forward to a meal. It is a steady pain in the belly and a longing for food. The person loses weight as the body's supplies of **fat** are used up. Then, the body begins turning the protein in muscle into energy. As their muscles shrink, the shape of their bones can be clearly seen through their skin. Their ability to fight disease is damaged and they often become sick. Children stop growing and are very weak and tired. The development of their brains is also affected and they find it difficult to learn or think clearly.

Growing Taller

Many people from **developing countries** move to **developed countries** in order to earn more money. Their children then eat a healthier diet than they did when they were young. When the children grow up, they are often taller than their parents.

How tall you grow does not depend entirely on the amount of protein in your diet. These two children are the same age, but they are different heights. Heredity, or the passing on of certain traits from a parent to a child, can also affect how tall you are.

Treating Marasmus

People who have been starved of food cannot eat large amounts at first. They need to be fed with regular small amounts of water that contains sugar and salts. Gradually, they may begin to drink milk and then eat solid food.

Mothers and babies

Pregnant women, mothers, and young children are most at risk in a famine. Pregnant women have to consume more than they usually do to support the developing child inside of them. Once born, babies may feed on their mothers' breast milk for up to two years. Mothers need extra protein so that their babies can grow.

Causes of famine

Although the world can produce enough food to feed everyone, millions of people starve to death every year. Most often, people starve because they cannot afford to buy food, especially food that contains protein.

areas most likely to suffer from famine

These are the parts of the world that are most likely to suffer from famine. They include countries where people are extremely poor and where there is often no rain for months or years.

Healthy Eating

Your diet is all the food you normally eat. To have a healthful diet, you need to eat a variety of different kinds of food to give you all the **nutrients** you need, that is, all the **carbohydrates, fats, vitamins,** and **minerals** as well as the protein you need. Of course, you do not think of food as proteins, carbohydrates, and other nutrients, but as bread, fruit, cheese, and so on. The Food Guide Pyramid divides foods into six main groups and shows how much you should eat from each group every day.

Bread, cereal, rice, and pasta

This group of foods is rich in carbohydrates. You should eat more servings from this group than from any other group—six to eleven servings each day. These foods also contain vegetable protein and some important vitamins and minerals.

Vegetables

Vegetables are rich in vitamins and minerals. You should eat three to five servings of vegetables every day. Vegetables also contain plenty of **fiber.** Fiber is not a nutrient, but it helps keep your digestive system working well.

Fruits

Fruits, like vegetables, are an excellent source of vitamins, minerals, and fiber. They are also an important source of carbohydrates. You should eat two to four servings from this group. An example of a serving of fruit would be a medium-sized pear or three-fourths of a cup (180 milliliters) of orange juice.

Meat, poultry, fish, dry beans, eggs, and nuts

These foods are rich in proteins, and you should eat two to three servings of food from this group every day. Some meats, such as pork, have more fat in them than, for example, chicken or beans. Adults and children who are overweight should try to choose foods that are low in fat from this group.

Milk, yogurt, and cheese

Foods in this group are rich in protein and other nutrients but can be high in fat. Adults and children who are overweight should look for low-fat varieties. Whole milk has the highest fat content. Low-fat (2% milk) milk has about half the fat of whole milk. Skim milk has almost no fat.

Fats, oils, and sweets

At the top of the pyramid are fats, oils, and sweets. We should eat these sparingly, or not often and in small amounts. These foods include butter, potato chips, cookies, and chocolate. These foods have few nutrients and are high in fats that lead to weight gain and health problems.

Tips for Healthy Eating
- Eat a wide variety of foods.
- Pay attention to serving sizes.
- Avoid eating too many fats, oils, and sweets.

The Food Guide Pyramid shown below was created to give you an idea of what to eat each day to maintain a healthful diet.

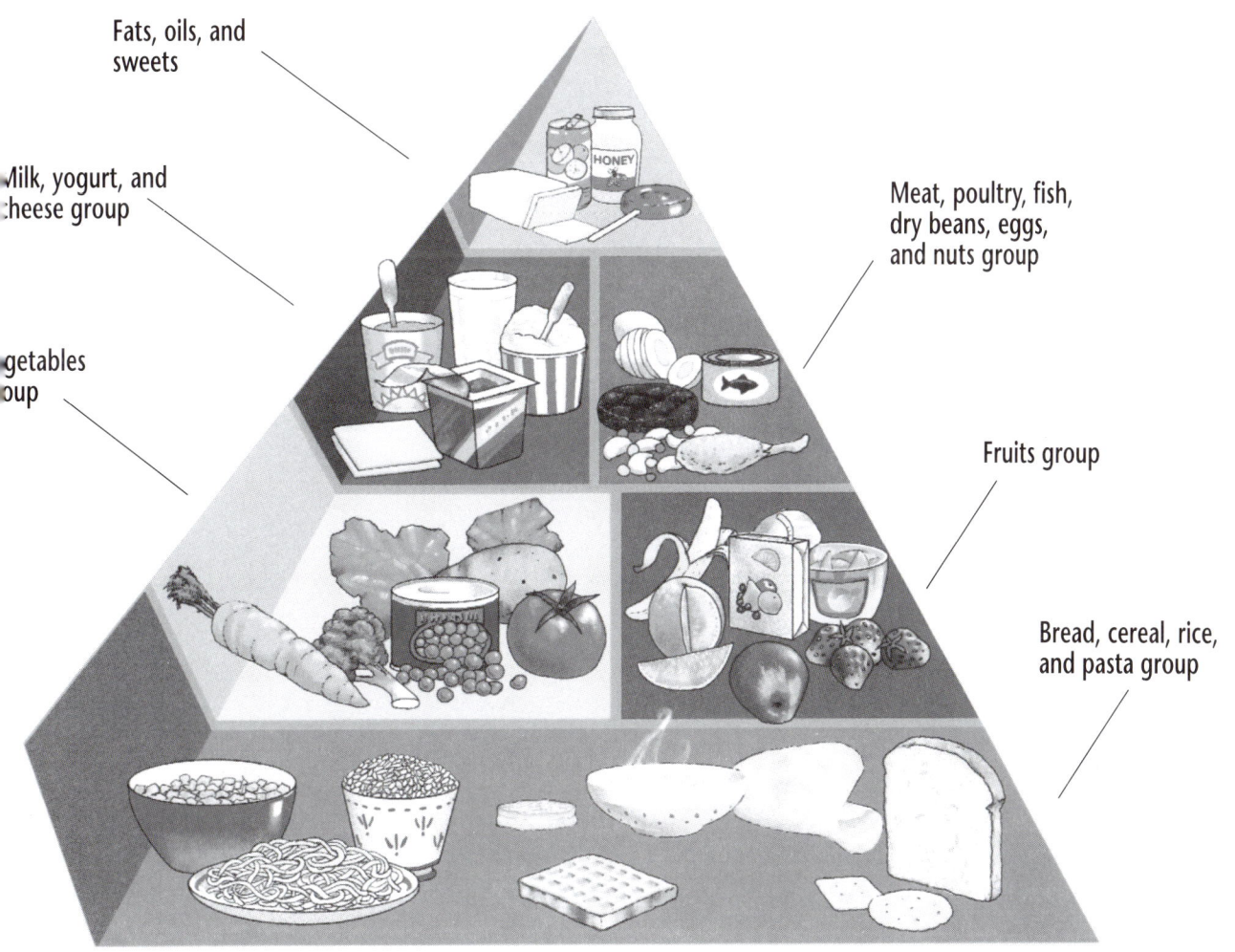

Fats, oils, and sweets

Milk, yogurt, and cheese group

Vegetables group

Meat, poultry, fish, dry beans, eggs, and nuts group

Fruits group

Bread, cereal, rice, and pasta group

HONEY

Different Kinds of Diets

People who live in different parts of the world often eat foods we may not be as familiar with. Many people follow diets that purposely leave out one or several kinds of food for religious reasons. For example, Muslims and Jews do not eat pork. Hindus do not eat beef. Vegetarians do not include meat in their diets, and vegans do not eat anything that comes from animals at all.

International food

The traditional foods of countries such as Mexico, China, India, and Japan are quite healthful. Many traditional dishes from these countries have become popular in the United States and other countries. Dishes from these countries sometimes use strong spices and special ingredients such as chili peppers, coconut milk, and seaweed. But most of them are based on foods that are familiar to everyone—rice, vegetables, chicken, and fish. Food cooked in different ways can be tasty and nutritious.

A vegetarian Indian meal can include a wide variety of different foods, including lentils, rice, bread, and yogurt. It provides all the protein you need.

Vegetarians

Vegetarians are people who do not eat meat or fish. They do, however, sometimes eat animal protein in the form of eggs and dairy products, such as milk and cheese. A vegetarian diet is perfectly healthful, particularly if each meal includes more than one source of protein. Vegetarians can be sure of getting more than one source of protein by eating **legumes,** such as soybeans or lentils, with rice or bread. They could also eat an animal protein, such as cheese or yogurt, with a vegetable protein. A vegetarian diet can be more healthful than the diet of a meat eater, because it is likely to contain less animal **fat.** It is not healthy to eat a lot of animal fat.

Vegans

Vegans do not eat any food that comes from animals. They eat only plants. Obviously, they do not eat meat or fish. They also do not eat milk, cheese, eggs, yogurt, or honey. Many packaged and processed foods contain milk powder or some other animal product, so vegans avoid these products as well. A vegan diet can provide all the protein a person needs. Like vegetarians, vegans must make sure to eat protein from more than one source at each meal. Vegans, however, have to plan their meals carefully to make sure they get all the **vitamins** and **minerals** they need.

Athletes

Professional athletes may need some extra protein to build their muscles. But most people eat more protein than they need. People who play sports for fun may need to eat extra **carbohydrates** for **energy,** but not more protein.

A vegetarian would eat this meal of a baked potato and cheese, but a vegan would not.

Healthy Snacks

Most people have favorite snacks that they eat when they are hungry between meals. Food companies produce a huge variety of snack foods including different kinds of chips, candy bars, cookies, and crackers. Movie theaters sell buttered popcorn and candy for people to eat while they watch a movie. Eating too many of the snacks can be unhealthy because they contain a lot of sugar and a lot of **fat.** But many other snacks are tasty and healthful, too. Here are some snacks that are rich in protein without being unhealthful.

Nuts and seeds

Nuts are a good crunchy snack. Peanuts, cashews, almonds, and walnuts are filled with protein and other **nutrients.** Seeds, such as sunflower seeds and pumpkin seeds, also contain protein. One thing to remember is that nuts and seeds contain quite a bit of fat. So limit yourself to a small handful, for example, of about 28 peanuts or about 22 almonds.

Hummus, taramasalata, and tsatsiki are the names of some traditional dips. They are tasty and rich in protein. You can make your own tsatsiki by mixing chopped-up cucumber, onion, mint, and garlic into plain or natural yogurt.

Popcorn

Popcorn is a type of corn and a source of protein. The popcorn you buy at the movies is served with a lot of oil or butter and salt. It is easy and more healthful to pop your own at home. You put raw popcorn kernels in a pot with a little bit of oil and then put the lid on and turn on the burner. You can hear the corn hitting the lid and sides of the pot as it pops. Do not remove the lid until the kernels have stopped popping! While it is still hot, add a little salt for flavor.

Ask Before You Cook!

If you are going to make your own snacks, ask an adult first. If you have not cooked before, you may need an adult to help you. If you use the oven or stove, make sure you turn it off when you have finished.

A toasted cheese and tuna sandwich is a healthful snack. Sandwich toasters get very hot, so ask an adult to help you if you make a snack like this one.

Nutritional Information

The top table shows how much protein children of different ages and weights need. Pregnant women or women who are nursing babies may need more protein than other adults.

Daily requirements of protein

Age and Weight	Protein (per day)
Age 9, 65 pounds (30 kilograms)	1.02 ounces (29 grams)
Age 10, 75 pounds (34 kilograms)	1.20 ounces (34 grams)
Age 11, 85 pounds (39 kilograms)	1.34 ounces (38 grams)

The table below shows amounts of particular foods needed to get one serving from five of the food groups in the Food Guide Pyramid. Remember to go easy on foods from the fats, oils, and sweets group at the top of the of the pyramid.

Recommended servings and examples of servings from the Food Guide Pyramid

Food Group	Recommended Servings	Examples of One Serving
Bread, cereal, rice, and pasta group	6–11 servings	1 slice of bread 1/2 cup (100 grams) of cooked cereal, rice, or pasta
Vegetable group	3–5 servings	1 cup (200 grams) of raw, leafy vegetables 1/2 cup (100 grams) cooked vegetables
Fruit group	2–4 servings	1 apple, banana, or orange 3/4 cup (180 milliliters) of fruit juice
Meat, poultry, fish, dry beans, eggs, and nuts group	2–3 servings	2–3 ounces (57–85 grams) of cooked lean meat, poultry, or fish
Milk, yogurt, and cheese group	2–3 servings	1 cup (230 milliliters) of milk or yogurt 1 1/2 ounces (40 grams) of natural cheese

This table shows how much protein is contained in 3.5 ounces (100 grams) of a variety of foods. It allows you to compare different foods and to figure out how much protein is in foods you eat.

Average amount of protein in some foods

Food	Protein		Food	Protein	
Milk and dairy foods			**Meat**		
Whole milk	0.11 oz	3.2g	Bologna	0.41 oz	11.6 g
2% milk	0.12 oz	3.3 g	Lean ground beef	0.62 oz	17.5 g
Fruit yogurt	0.18 oz	5.1 g	T-bone steak	0.59 oz	16.6 g
Cheddar cheese	0.90 oz	25.5 g	Beef liver	0.69 oz	19.7 g
Hard-boiled egg	0.44 oz	12.5 g	Salami	0.46 oz	13.0 g
			Pork sausage	0.47 oz	13.3 g
Breads and grains			Roast chicken	0.87 oz	24.8 g
White bread	0.30 oz	8.4 g	Roast turkey	0.99 oz	28 g
Whole wheat bread	0.32 oz	9.2 g			
Cornflakes	0.28 oz	7.9 g	**Legumes**		
Cooked spaghetti	0.13 oz	3.6 g	Canned baked beans	0.18 oz	5.2 g
Boiled rice	0.09 oz	2.6 g	Tofu (from soybeans)	0.29 oz	8.1 g
			Cooked lentils	0.27 oz	7.6 g
Fish					
Grilled fish sticks	0.53 oz	15.1 g	**Snacks**		
Boiled shrimp	0.70 oz	19.8 g	Chocolate cookies	0.20 oz	5.7 g
Canned sardines	0.84 oz	23.7 g	Peanuts	0.86 oz	24.5 g
Canned tuna	0.83 oz	23.5 g	Hummus	0.27 oz	7.6 g

grams = g   ounces = oz

How Much Protein?

A pizza that weighs 25 ounces (720 grams) has 2.15 ounces (61 grams) of protein. If four nine-year-old boys weighing 65 pounds (29 kilograms) each share it, each boy will get about half the protein he needs for the day. But for four sixteen-year-old boys weighing 150 pounds (68 kilograms) each, the pizza will provide only about a fourth of the protein they need.

Glossary

acid   chemical compound that aids in digestion

amino acid   smaller unit or building block of proteins. Different amino acids combine together to form a protein.

anaphylactic shock   severe allergic reaction in which a person's body reacts so strongly that he or she could die

anorexia   eating disorder in which a person eats very little food and becomes extremely thin

antibody   cell that attacks certain kinds of bacteria, viruses, or poisons

artery   tube that carries blood from the heart to different parts of the body

bacteria   microscopic living things. Some are helpful, like those in our intestines, but some can cause disease.

bone marrow   jellylike substance in the center of some bones, where red and white blood cells are made

bulimia   eating disorder in which a person eats a lot of food and then throws up

carbohydrate   substance in food that the body uses to provide energy. Foods rich in carbohydrates include bread, rice, potatoes, and sugar.

carbon   invisible gas in the air and one of the most common elements. Carbon is one of the elements that makes up an amino acid.

cell   smallest unit of a plant or animal

chemical bond   something that joins together two molecules or atoms

chemical reaction   when two or more chemicals react together to produce a change

chyme   mushy liquid that passes from the stomach to the small intestine

cytoplasm   all the material inside a cell except the nucleus

developed country   wealthy country that has well-established industries and services

developing country   poorer country that does not have well-established industries or services

energy   ability to do work or to make something happen

enzyme   substance that helps a chemical reaction take place faster

esophagus   tube through which food travels from the mouth to the stomach

essential amino acid   one of nine amino acids that have to be acquired from food because the body cannot make them

famine lack of food that affects most people in a region. Famine leads to starvation.

fat substance found in a wide range of foods. The body can change fat into energy. Fat is stored by the body in a layer below the skin.

fermented slowly broken down by fungi, such as yeast, to form a new substance

fertilizer chemical or natural substance added to soil to provide nutrients for plant growth

fiber substance found in plants that cannot be digested by the human body

fungus simple living thing such as a mushroom, yeast, or mold

gland part of human body that produces fluid that the body either uses or gives off as waste

glucose simple form of sugar that is broken down from carbohydrate food during digestion

glycogen substance made from glucose that is stored in the liver and muscles following absorption

hormone substance made by different glands in the body that affects or controls certain organs, cells, or tissues

hydrochloric acid strong acid made when a gas made up of hydrogen and chlorine is dissolved in water

hydrogen invisible gas in the air. Hydrogen is an element that combines with other substances to form, for example, water, sugar, proteins, and fats.

keratin kind of protein that the hair, nails, and outer layer of skin are made of

legume type of food that includes lentils, peas, and beans. Legumes are the seeds of certain plants and are rich in protein.

liver organ in the body that plays a role in digestion. It makes bile and helps clean the blood. People also eat beef and chicken livers, which are a rich source of protein, vitamins, and minerals.

mineral nutrient found in foods that the body needs to stay healthy

mitosis process by which a cell divides to form two new identical cells

molecule smallest unit of a substance that is still that same substance and still has the same properties as the substance

nitrogen invisible gas that is the main gas in the air. Nitrogen is one of the elements that make up amino acids.

nucleus part of the cell that controls all the cell's activities

nutrient substance found in foods that helps the body grow and stay healthy. Proteins, carbohydrates, fats, vitamins, and minerals are all nutrients.

organ body part that has a particular job to do. An eye is an example of an organ.

oxygen gas present in the air and used by the body. Oxygen is one of the most common elements and is used by the body to make amino acids.

pancreas gland that produces various digestive juices that flow through a tube into the small intestine

pepsin enzyme in the stomach that starts to break down proteins

peptide simple chain of amino acids

peptide bond chemical bond that joins peptides together

phosphorus simple substance that is needed for plants to grow

puberty time of life when a child's body develops into an adult's body

red meat meat such as beef, lamb, or pork

ribosomes particles in the cells of the body that build up proteins from amino acids

saliva watery liquid made by glands in the mouth and the inside of the cheeks

sequence several things in a particular order

sulfur one of the elements that forms part of most proteins

tissue material made up of cells that forms a part of an animal or of a plant

urea substance formed from waste material in the liver.

villus tiny, fingerlike extensions in the small intestine through which digested food and water are absorbed. More than one villus are called villi.

virus tiny, nonliving thing inside your body that can make you sick

vitamin nutrient needed by the body in small amounts

yeast type of fungus. Yeast is used to make bread dough rise.

Further Reading

Ballard, Carol. *The Digestive System.* Chicago: Heinemann Library, 2002.

Brownlie, Ali. *Why Are People Vegetarian?* Austin, Tex.: Raintree Publishers, 2002.

D'Amico, Joan, and Karen Eich Drummond. *The Healthy Body Cookbook.* Hoboken, N.J.: John Wiley & Sons, 1999.

Gregson, Susan R. *Healthy Eating.* Mankato, Minn.: Capstone Press, 2000.

Hardie, Jackie. *Blood and Circulation.* Chicago: Heinemann Library, 1998.

Kalbacken, Joan. *The Food Pyramid.* Danbury, Conn.: Children's Press, 1998.

Rondeau, Amanda. *Proteins Are Powerful.* Edina, Minn.: ABDO Publishing, 2002.

Royston, Angela. *Eating and Digestion.* Chicago: Heinemann Library, 1998.

Weintraub, Aileen. *Everything You Need to Know about Eating Smart.* New York: Rosen Publishing, 2000.

Westcott, Patsy. *Diet and Nutrition.* Austin, Tex.: Raintree Publishers, 2000.

Index